We're making a coat
    for the crocodile.
It has to be
    his special style.

First we need to measure
his height.
We must be sure
to get it right.

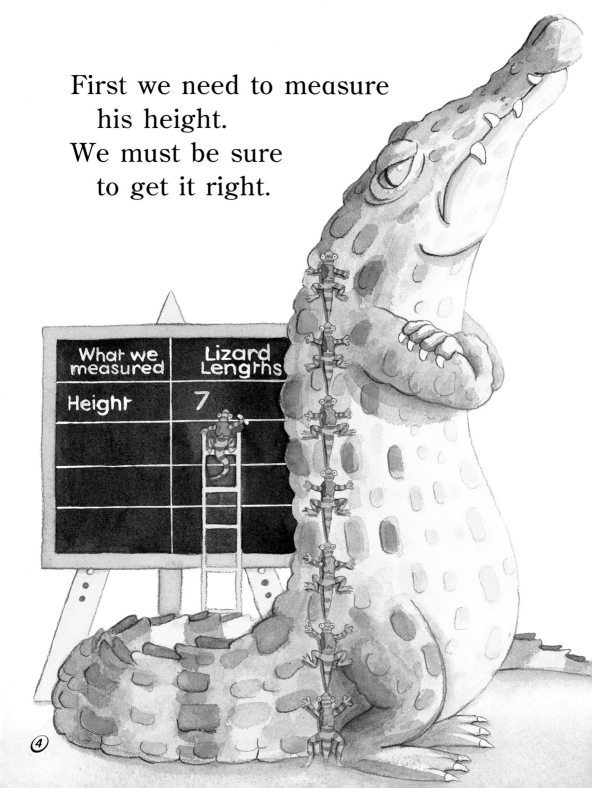

What we measured	Lizard Lengths
Height	7

④

Now we'll measure
    around his chest —
he puffs it out
    to look his best.

Now his arms —
    don't get those wrong —
Croc doesn't like his sleeves
    too long.

What we measured	Lizard Lengths
Height	7
Chest	6
Arm	3

That just leaves
    his scaly neck.
We'll measure it twice
    to double check.

What we measured	Lizard Lengths
Height	7
Chest	6
Arm	3
Neck	4

There are two more on that side.

At last the measuring
is all done,
but our hard work
has just begun.

What we measured	Lizard Length
Height	7
Chest	6
Arm	3
Neck	4

Now we have to cut
    and sew,
and Croc's impatient —
    we can't be slow.

We've finished the coat
for the crocodile,
and he likes it, too —
just look at him smile!

The Crocodile's Coat

Written by Calvin Irons

Illustrated by Peter Shaw

Stories about measurement...

The Crocodile's Coat
length

Fishy Scales
weight

The Jumping Contest
length

The Bears' Breakfast
capacity

Baby Bear's Quilt
area

The New Fence
length

These six stories provide a continuum of learning development.
They should be used in sequence to develop important concepts
and skills during the first four years of school.

© 1999 this edition
Mimosa Publications Pty Ltd

Published in the United Kingdom by KINGSCOURT PUBLISHING
Published in the United States of America by MIMOSA EDUCATION
Published in Canada by PRENTICE HALL GINN CANADA
Published in New Zealand by SHORTLAND PUBLICATIONS
Published in Australia by SHORTLAND-MIMOSA

Printed in Hong Kong   6 5 4 3 2

ISBN 0 7327 2698 0

8180

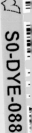